SCIENCE QUESTIONS

HOW DO PUPPIES GET THEIR COLORS?

by Elizabeth Andrews

Cody Koala
An Imprint of Pop!
popbooksonline.com

abdobooks.com
Published by Pop!, a division of ABDO, PO Box 398166, Minneapolis, Minnesota 55439.

Printed in the United States of America, North Mankato, Minnesota

102021
012022

THIS BOOK CONTAINS RECYCLED MATERIALS

Cover Photo: Shutterstock Images
Interior Photos: Shutterstock Images, 1, 5 (top, bottom), 6, 7, 9, 10, 13, 14, 17, 18, 21 (top, bottom left), Wikimedia, 17 (bottom middle)

Editor: Grace Hansen
Series Designer: Laura Graphenteen

Library of Congress Control Number: 2021942296

Publisher's Cataloging-in-Publication Data
Names: Andrews, Elizabeth, author.
Title: How do puppies get their colors? / by Elizabeth Andrews
Description: Minneapolis, Minnesota : Pop!, 2022 | Series: Science questions | Includes online resources and index.
Identifiers: ISBN 9781098241087 (lib. bdg.) | ISBN 9781098241780 (ebook)
Subjects: LCSH: Puppies--Juvenile literature. | Animals--Color--Juvenile literature. | Puppies--Breeding--Juvenile literature. | Children's questions and answers--Juvenile literature.
Classification: DDC 500--dc23

Hello! My name is

Cody Koala

Pop open this book and you'll find QR codes like this one, loaded with information, so you can learn even more!

Scan this code* and others like it while you read, or visit the website below to make this book pop.

popbooksonline.com/puppy-colors

*Scanning QR codes requires a web-enabled smart device with a QR code reader app and a camera.

Table of Contents

Chapter 1

Pick of the Litter

There is nothing quite as cute as a litter of puppies. A litter is the group of puppies that a female dog has at one time.

Watch a video here!

Picture a litter of dachshunds (dahks-hunds). Most of the young pups have black coats, like their parents. But there are two that are brown.

Puppies from the same litter can look different from one another. Genes are the reason for these differences.

Litters can have anywhere from 1–12 puppies in them!

Chapter 2

What Are Genes?

Genes are instructions that tell the body how to look and act. Every person and animal has two copies of each gene, one from each parent. Genes are made of **DNA**.

Learn more here!

If all DNA inside a single chromosome was unwrapped, it would stretch 6 feet (1.8 m)!

Genes are found in the middle of **cells**, on chromosomes. Chromosomes are tightly packed clumps of DNA. People often have 23 pairs of chromosomes in each cell. Most of them are shaped like an X.

Chapter 3

What Are Alleles?

The **combination** of genes a puppy gets from its parents decides the puppy's physical **traits**. Every puppy has two copies of each gene. Alleles are forms of the same gene.

Learn more here!

Capital and lowercase letters stand for alleles. Think of the dachshund litter again. Both parents have black coats. They also each have *Bb* alleles. Capital *B* is for black. Lowercase *b* is for brown. This means each parent could pass either trait on to each puppy.

Chapter 4

Punnett Square

Reginald Punnett created a simple way to predict how **traits** may get passed on. It is called the Punnett square. Each section in the square shows an allele that a puppy could get.

Reginald Punnett

Complete an activity here!

DACHSHUND PUPPY COAT COLOR

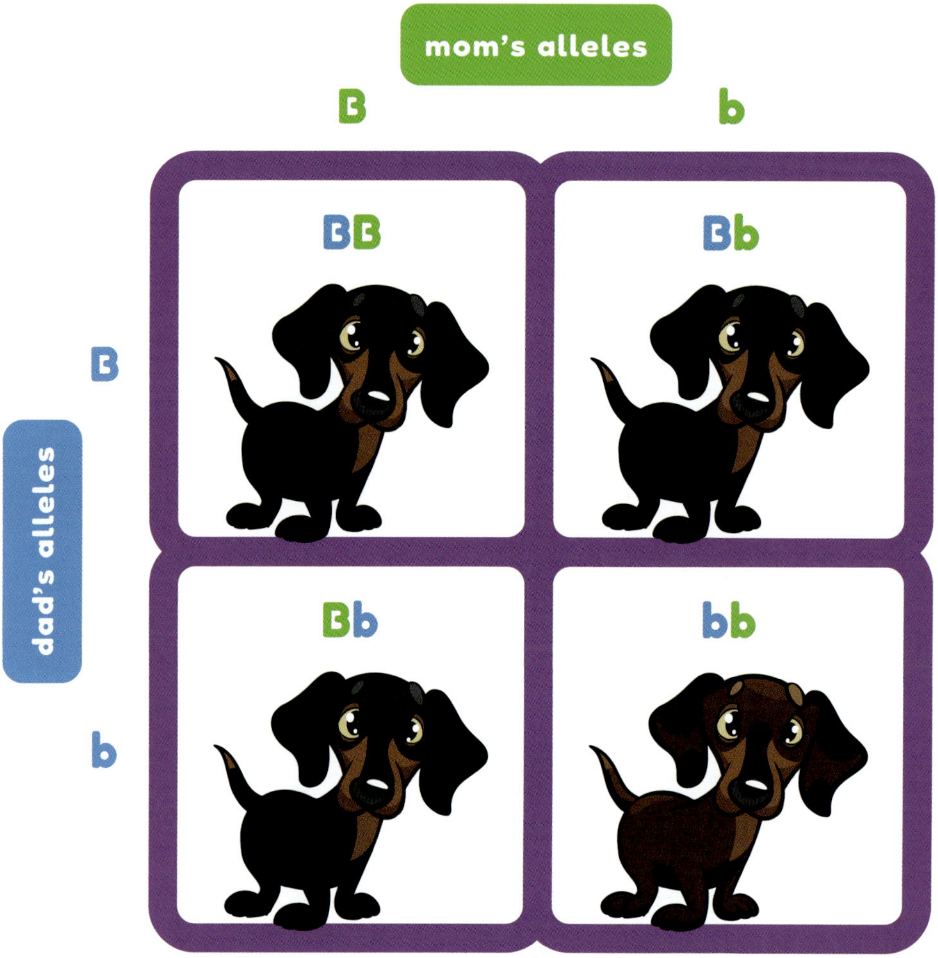

UPPERCASE LETTER: **dominant** gene
lowercase letter: **recessive** gene

If a puppy **inherits** even one uppercase letter, it will have a black coat. In this litter, one *bb* **combination** is possible. This means the puppies only have a one-in-four chance to have a brown coat. They are more likely to have black coats.

Humans have genes too. They inherit different ones from each parent just like puppies. This is why people don't all look the same. Everyone is made of a special combination of genes!

Identical twins look exactly the same because they inherited matching genes!

Making Connections

Text-to-Self

Have you ever seen a litter of puppies? Did they look alike or different from one another?

Text-to-Text

Have you read any other books about genes? If so, did you learn anything different in that book?

Text-to-World

Do you think it's good that everyone looks different? What would the world be like if people all looked the same?

Glossary

cell – the smallest part of a living thing.

combination – a mixture of different things.

DNA – short for deoxyribonucleic acid. A material in the body that helps determine what features a living thing will have.

dominant – relating to the controlling gene in an allele.

inherit – to receive from a parent.

recessive – relating to a gene that will not be expressed if paired with a dominant gene in an allele.

trait – a quality or feature of something.

Index

Online Resources

popbooksonline.com

Thanks for reading this Cody Koala book!

Scan this code* and others like it in this book, or visit the website below to make this book pop!

popbooksonline.com/puppy-colors

*Scanning QR codes requires a web-enabled smart device with a QR code reader app and a camera.